On the Move

BOATS

Ursula Pang

PowerKiDS press.

PK Beginners

Look at all the boats!

Boats move on the water.

We wear life vests
on the boat.
It's safe!

LIFE VEST

We use oars to move the boat.

OAR

Sailboats have sails.

SAIL

Wind makes sailboats go!

Motorboats use motors to move.

MOTOR

Motorboats can
be fast!

Some people use boats for fishing.

Some boats are
really big!

Would you like to take a ride on a boat?

Published in 2025 by The Rosen Publishing Group, Inc.
2544 Clinton Street, Buffalo, NY 14224

First Edition

Editor: Greg Roza
Book Design: Rachel Rising

Photo Credits: Cover, p. 1 Nata-Lia/Shutterstock.com; p. 3 Don Donelson/Shutterstock.com; p. 5 BlueOrange Studio/Shutterstock.com; p. 7 New Africa/Shutterstock.com; p. 9 SaskiaAcht/Shutterstock.com; p. 11 Netfalls Remy Musser/Shutterstock.com; p. 13 Vinnikava Viktoryia/Shutterstock.com; p. 15 Annado/Shutterstock.com; p. 17 javarman/Shutterstock.com; p. 19 Alesia Verem/Shutterstock.com; p. 21 Aun Photographer/Shutterstock.com; p. 23 Denis Belitsky/Shutterstock.com.

Library of Congress Cataloging-in-Publication Data

Names: Pang, Ursula, author.
Title: Boats / Ursula Pang.
Description: [Buffalo] : PowerKids Press, [2024] | Series: On the move
Identifiers: LCCN 2023050561 (print) | LCCN 2023050562 (ebook) | ISBN 9781499444575 (library binding) | ISBN 9781499444568 (paperback) | ISBN 9781499444582 (ebook)
Subjects: LCSH: Boats–Juvenile literature.
Classification: LCC VM150 .P364 2024 (print) | LCC VM150 (ebook) | DDC 623.82–dc23/eng/20231103
LC record available at https://lccn.loc.gov/2023050561
LC ebook record available at https://lccn.loc.gov/2023050562

Manufactured in the United States of America